About Being able to Look

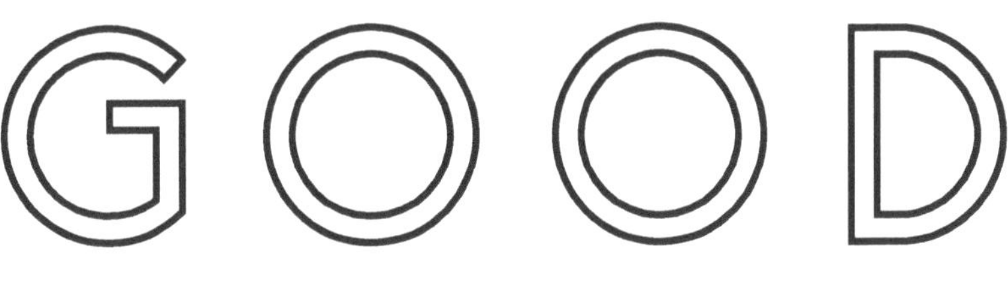

in a Burlap Sack

IDA TOMSHINSKY

ISBN: Softcover 978-1-4797-6120-3

This is a work of fiction. Names, characters, places and incidents either are the product of the author's imagination or are used fictitiously, and any resemblance to any actual persons, living or dead, events, or locales is entirely coincidental.

To order additional copies of this book, contact:
Xlibris Corporation
1-888-795-4274
www.Xlibris.com
Orders@Xlibris.com

About Being able to Look GOOD in a Burlap Sack

Burlap is a jute product with a long history, eco-friendly present and promising future. *"Burlap is probably as old as Moses,"* stated costume designer, Emilio Sosa from "Project Runway" TV show. Fashion Designer, Michael Kors thinks that a burlap cloth could be 'hot' and 'edgy.' Nina Garcia likes the three-big-F's idea of using burlap fabric in fashion design because it *"comes out from the farm, to the future, and it's fabulous."* Burlap fabric is perhaps most familiar as sacking material, if not from potato or rice sacks then from childhood sack races. The rough, sometimes hairy texture and familiar brown color associates with "gunny" sacks and belies in burlap's versatility. In addition to its industrial and agricultural uses, burlap has applications in craft projects and home decors.

Burlap predominately made in India, Bangladesh and Pakistan. Burlap is a rough-textured, loosely woven fabric that is usually made from jute fibers. Jute is a natural plant fiber and a member of the hibiscus family. The fibers come from the inner part of the plant's stem. Burlap is also known as *hessian* and *jute hessian*. Burlap fabric characteristics include heavy, strong and hard-wearing. Jute fibers are naturally resistant to stretching, and the porous weave of burlap makes it breathable. Burlap varies in its texture and appearance, from hairier, heavier and rougher fabrics for industrial use to more smoother, more tightly woven burlap sold in fabric stores. The latter is available in a wide range of colors and resembles heavy linen. Jute burlap is hundred percent biodegradable, naturally resistant to fire, flexible and absorbent. It readily accepts dye and stands up to heavy handling.

Burlap is familiar as a sacking material

Burlap is often used to make potato, grain, onion or rice sacks. Agricultural uses include packaging for bales of wool, and wrapping tree roots and trunks for protection during transit. It is used as a carpet backing and as a cover for wet cement to slow down the drying time. Burlap can be laminated, bleached or dyed, and sometimes it is treated with chemicals to be rot-resistant. As a decorative fabric for home sewing projects, burlap comes in a range of colors and offers a rustic, hand-loomed look that complements a country-style decor. When evenly woven, burlap works well as a backing fabric for cross stitch and needlepoint.

Burlap fabric was originally used for curtains in interior design, but now has a wide range of uses from the decorative to the utilitarian. With its strength and low cost, a tremendous amount of burlap is used in the construction industry, where it is used to cure concrete. And with its water ability to disintegrate over time, making it eco-friendly, burlap is a favorite fabric of gardeners. Burlap is still used for curtains and wallpaper, in theatrical productions, and in many kinds of craft projects. The utility burlap is an economical, quality product without the stamps and markings of inferior grades, and the color burlap is crisp and bright-terrific for a wide range of uses.

- Practical note for working with burlap fabric:

  When people use burlap fabric in a sewing project, they have to be aware that its tendency to fray at the edges means to use bound seam finishes. Use bias tape to encase the cut edges of a seam or use an over locking stitch along the seam edges. This will enhance the finished appearance of the project and prevent unraveling. Painter's tape will temporarily prevent unraveling along the edges. It is easy to use burlap fabric for needlework, align the straight and cross grains before cutting the backing piece. To do this, will need to pull one of the threads running along the edge of the fabric, gently teasing it out. Use the line of the pulled thread to cut a perfectly straight square. The rough texture and heaviness of burlap limits its use somewhat. For example, it is not commonly used for garment making, but it does have a place in fashion and home decor, though, as in the form of casual purses and simple curtains. While the tendency to fray at the edges when cut makes burlap unsuitable for some projects and requires special sewing techniques, it can also be used for decorative effect, for example, by creating a fringe.

Burlap comes from minor natural vegetable fibers that are generally composed mainly of cellulose: examples include cotton, jute, flax, ramie, sisal, and hemp. Cellulose fibers serve in the manufacture of paper and cloth.

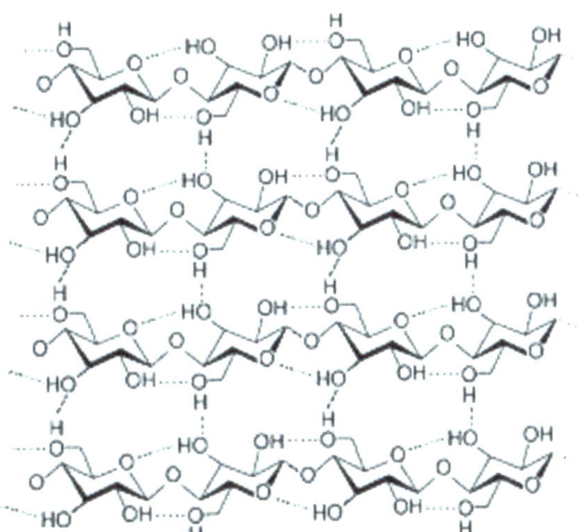

Fig. 1 Illustration of a strand of cellulose showing the hydrogen bonds (dashes) within and between cellulose molecules in the family of the vegetable fibers

Minor natural vegetable fibers can be further categorized into the following description:

Category	Description
Seed fiber	Fibers collected from seeds or seed cases. e.g. cotton and kapok
Leaf fiber	Fibers collected from leaves. e.g. fique, sisal, banana and agave.
Bast fiber	Fibers are collected from the skin or bast surrounding the stem of their respective plant. These fibers have higher tensile strength than other fibers. Therefore, these fibers are used for durable yarn, fabric, packaging, and paper. Some examples are flax, jute, kenaf, industrial hemp, ramie, rattan, and vine fibers.
Fruit fiber	Fibers are collected from the fruit of the plant, e.g. coconut (coir) fiber.
Stalk fiber	Fibers are actually the stalks of the plant. E.g. straws of wheat, rice, barley, and other crops including bamboo and grass. Tree wood is also such a fiber.

The most used minor natural vegetable fibers are cotton, flax (linen) and hemp, although sisal, jute, kenaf, bamboo and coconut are also widely used.

What is fiber? Fiber comes in different classes of hair-like materials that are similar to pieces of tread. They can be spun into filaments, treads or rope. They can be used as a component of composite materials. They can also be matted into sheets to make products such as paper or felt. Fibers are of three types: natural fiber, cellulose fiber, and synthetic fiber. The earliest evidence for humans using fibers is the discovery of wool and dyed flax fibers found in a prehistoric cave in the Republic of Georgia that date back to 36,000 B.C.

The natural textile fibers are divided into three major categories (1) animal, (2) vegetable, (3) mineral. Animal fibers contain animal wool or hair, silk, and avian fiber. A good example of the natural mineral fiber is the asbestos. Note on natural textile fabrics:

- Animal hair (wool or hairs): Fiber or wool taken from animals or hairy mammals comprise proteins such as collagen and keratin that includes sheep's wool, goat hair (cashmere, mohair), alpaca hair, horse hair, etc.
- Silk fiber: Fiber collected from dried saliva of insects during the preparation of cocoons.

- Avian fiber: Fibers from birds such as feather fiber.

Burlap comes from the bast family of fibers. Note on bast family of fibers:

- Bast fibers are a family of vegetable fibers which run the length of the plant stem. They are fund in the inner bark of plants.

Flax fibers, the source of linen yarns and cloth, are the most important. Burlap is in the minor category of the bast fibers such as ramie or rhea, jute, and hemp.

Burlap is a material that is made out of the fibers of the jute plant. The material commonly called burlap in the United States is known as "hessian" in parts of Europe because it was used in the uniform of soldiers from the state of Hesse. In some places, a bag made from the burlap material is called a "gunny sack." And, of course, burlap is this rough material that makes up the tobacco, sugar, and potato bags strong and last for years. It's also sometimes blended with other fibers to make yarn, rope, cordage, nets and other similar products.

The texture of burlap comes from the fact that it is made from the skin of the jute plant – one of the least expensive textile crops in the world, and one of the strongest. Jute fibers have a high content of cellulose, a major component of plant stalks, as well as lignin, a constituent of wood. The result is a textile fiber that is half cloth and half wood with heat resistant, and quality of easiness to dye reflects on production and manufacturing. Jute cultivation requires standing water, so most production comes from Bangladesh and India, regions that benefit from a heavy monsoon season. After cotton and linen, it is the third most important vegetable fiber in the world.

Left to right: Jute plant – botanical drawing; jute field; jute leaves; jute flower

As it turns out, jute is a very versatile material. In clothing, the lignin component makes for an itchy material, but also could be separated into very fine threads when it has been used to make imitations of silk. The strength and low cost of jute fiber makes it ideal for industrial applications. The fact that it's entirely natural and biodegradable makes it a popular choice for uses where these qualities are desirable, such as in covering and binding plant roots for transportation and replanting, or to prevent erosion. Blended with other materials, jute fibers are woven into curtains, chair coverings, carpets, area rugs, canvas, and even, the backing for linoleum. A relatively new application for jute, capitalizing on its lignin content, is as a woody pulp for making paper and pressboard furniture. It has notable potential as a component in lightweight automobile exterior panels that resist side impact.

The plant has an antioxidant activity with a significant α-tocopherol equivalent to Vitamin E. In the English language, the leaves are called Jew's mallow. High in beta-carotene, iron, calcium and Vitamin C, and with significant antioxidant activity, the leaves are eaten alone or as part of a soup in many parts of the world. In West Africa and Egypt, in fact, jute is a key ingredient of the national dish. Mulukhiyah, Mulukhiyya or Malukhiyah is Arabic for leaves of the jute plant or related Corchorus species. They are of the brand of Corchorus, an herb in the Linden family. The green vegetable is commonly used in Middle-Eastern, mainly Egyptian cuisine, and is also used in some Far-East oriental dishes in countries such as Japan. It is a rather bitter herb with a natural thickening agent. It is the main ingredient of a popular Egyptian dish by the same name.

Let's turn to the world's national culinary books for a moment! Malukhiyah is prepared in various styles: the original Egyptian style wherein the mallow leaves are very finely chopped, with ingredients such as garlic and coriander added to give it a characteristic aromatic taste, or the Lebanese, Palestinian, Syrian and Jordanian style where the leaves are used whole. "Malukhiyah Stew" is served with rice, but usually enjoyed with chicken. "Chicken & Mallow leaves stew" is a popular dish in Syria; rabbit is substituted for chicken in the Egyptian version of the dish.

Malukhiyah has been known as a popular food in Egypt since the time of the Pharaohs, and later spread to the Levant. The leaf is a common food in many tropical West African countries. It is believed that the "drip tips" on the leaves serve to shed excess water from the leaf from the heavy rains in the tropics. It is called Kren-Kre in Sierra Leone, and is eaten in a palm oil sauce served with rice or cassava fufu, or is steamed and mixed into rice just before eating a non-palm oil sauce.

Jute plant (Corchorus olitorius)

Jute is a natural vegetable fiber with golden silky shine color and soft texture for what it also named "The Golden Fiber." It is the cheapest natural fiber next to cotton. The very fine jute treads are used for making silk imitation. Plus, it could be blended with other natural and synthetic fibers and used as dyes. The modern changing trend made jute "The Fiber-For-The Future" due to its versatility and durability. In the coming era, the mini plants and mills will produce more of these golden fibers in form of blended yarns.

There are 50-60 different Corchorus species, and all species are apparently high in fibrous [rubbery.] As usual, they are found in warm regions throughout the world, on all continents and on numerous tropical and sub-tropical islands. However, the centre of diversity and origin of the jute species appears to be Africa where the largest number of Corchorus species, as around thirty has been found with highest concentration of fiber reported from East and South Africa. Of the cultivated species of jute, there is an omnipresent of it in Indo-Mynamar and South China, from where it migrated into India and Bangladesh. In contrast, though many workers earlier held the view about this species being native to Sri Lanka, India and Kenya, it is now generally agreed to have finest jute originated in Africa and migrated to India and China via Egypt and Syria.

After a lot of debate, there seems to be an agreement that *White Jute* originated in the Indo-Burma region and *Tossa Jute* in Africa. *Kenaf* originated in Angola, in Africa, and *Roselle* originated in Sudan of Africa. (D. P. Singh, 1983). China is also considered as one of the places of origin of *Jute*. According to some scholars, some provinces of the southern parts of China are the secondary centers of origin of *Tossa* and *White Jute*. (Peikun Huang, 1992).

Jute cultivation requires specific climate and land. It requires early rains in March, May and June and intermittent rain and sunlight thereafter till August, temperature between 28°C and 35°C and humidity between 70% and 90%. This type of climate is available in areas between 30° Latitude North and South of the earth. *Kenaf* and *Roselle* grow almost throughout the world both in tropical and temperate areas.

"Jute is a bast fiber that comes chiefly from India, because the plant grows well in rich land, especially along tidal basins, through improved methods, financial aid, and greater acreage, has created its production. There has been some attend to raise jute along the Gulf of Mexico, but the cost of labor has been too high to warrant its cultivation." (Wingate, 1986)

The jute plant grows to a height of about twelve feet. It is cut off close to the ground when it is in flower. Like flax, it is stripped of its branches and leaves and it put through a retting process to loosen the fibers from the stalk. After they are separated from the outer bark, the fibers are dried and cleaned. The jute fibers are weaker than those of linen. The fibers are very short, but lustrous and sooth. Because jute is affected by chemical bleaches, it can never be made pure white. It is very durable and is very much weakened by dampness. It is affected by sunlight. Only neutral soaps containing no free alkaline should be used in laundry, as alkaline used in laundry weaken jute. Jute can be distinguished from linen and cotton if the fibers are stained with iodine and then concentrated sulfuric acid and glycerin are applied. After test, jute fibers remain yellow; cotton and linen turn blue. It made irresistible to use it for gunny sacks, burlap bags, cordage and rope, and binding and backing treads for rugs and carpets.

✓ Facts about Burlap

Potential Burlap Invention

- The English recognized the potential of the jute plant while occupying India. They noticed that it being used in paper and rope.

Exportation

- In the late part of the 18th century, the East India Company brought back 100 tons of jute to London.

First Mill

- The first jute mill was built in Calcutta, India, in 1855; but within five years, five more mills had been built.

India Prime Exporter

- Although other countries had mills to make burlap. By 1939, India had 68,000 looms working to make it the number-one producer of burlap in the world.

India and Pakistan

- When India and Pakistan separated into two nations, most of the jute plants were on Pakistani lands. India had to start over with its burlap production, and Pakistan pushed its burlap industry further.

Jute was used for many years by the people of India, but in small quantities for products such as rope and paper. When English traders saw the potential of the plant, they began to export large quantities of it. In 1793 the East India Company brought hundred tons to Britain. Some jute was brought to Dundee, Scotland, where eventually a process was designed to spin jute yarn in large volumes. Jute became an important export from the countries that produced it. In 1855, Calcutta became the site of the first jute mill in India, close to the source of the plant. Mills began to sprout up in that region of the world; and by 1869, there were five mills operating a total of almost 1,000 looms. When a way to make better grade burlap was invented India began to dominate the world market for that particular jute product. Other countries around the world began to manufacture jute, but India had as many as 68,000 looms going full tilts by 1939. When the Indian sub-continent was divided, India lost access to previous lands that were devoted to jute farming, as they were in Pakistan's territory. India was forced to produce its own jute plants, while Pakistan became a major player in the jute market.

Today, jute is grown in India and Bangladesh, which is what eastern Pakistan became. These countries dominate the jute production of the world, followed by nations like China, Myanmar, Brazil and Thailand. The strength of burlap is renowned, as it is hard to tear and can stand up to great pressure. Burlap is extremely weather resistant and can be dried over and over again after becoming moist. It is also available in many widths, weights and forms. Burlap is able to be colored, sewn, treated to protect against rotting and even laminated.

There are many uses for burlap in today's world besides being used for bags. It can be made into windbreaks to protect growing trees. It is also used to prevent erosion on hillsides, especially when planting new lawns. Burlap offers great protection from animals such as mice and rabbits to young seedling trees just being planted. Wool is often shipped in burlap and burlap is an important part of furniture making, giving support to the inner portions of chairs and couches. Burlap is described as a "breathable" fabric, meaning that it is resistant to condensation. Because the contents of a burlap bag will not be able to absorb moisture burlap has been used to make all types of sacks and bags for the purpose of shipping such goods as coffee. It is a durable fabric as well, perfect for the rigors that it must endure while containing goods that are shipped from port to port. Burlap, because of the all above considerations, is also used to protect cement and concrete that is in the process of setting.

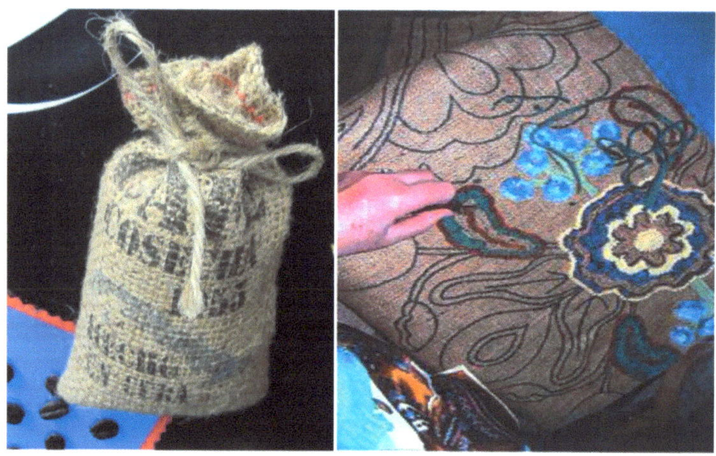

Left to right: burlap miniature sack and burlap base for the rug

Hessian in Europe or burlap as it called in the United States is a woven fabric usually made of the jute plant or sisal fibers. Gunny cloth is similar. Hessian, a dense woven fabric, has been historically produced as a coarse fabric. The name "burlap" appears to be of unknown origin. Due to its coarse texture, it is not commonly used in modern apparel. However, this roughness gave it a use in a religious context for *mortification of the flesh*, where individuals may wear an abrasive shirt called a cilice or "hair-shirt" and in the wearing of "sackcloth" on Ash Wednesday. Owning to its durability, open weave, naturally non-shiny refraction, and fuzzy texture, ghillie suits for 3-D camouflage are often made of hessian. Note on the ghillie suits:

- A ghillie suit, also known as a wookie suit, yowie suit, or camo tent is a type of camouflage clothing designed to resemble heavy foliage

It was also a popular material for camouflage scrim on combat helmets of World War II. Until the advent of the plastic "leafy" multi-color net system following the Vietnam War, burlap scrim was also woven onto shrimp and fish netting to create large-scale military camouflage netting. Hessian is also commonly used for making sandbags, empty hessian sacks that, when filled with sand, are used for flood mitigation when building temporary embankments against floodwaters or field fortifications. 100% natural burlap possibly is the most versatile home and garden product ever. 100% natural burlap had hundreds of uses both indoors and out. Burlap protects seedlings, controls erosion, and serves as a shade cloth, wind screen or camouflage. Indoors, its uses are limited only by creative imagination. Burlap's rich texture makes it ideal for craft projects hundreds of uses at a very low price.

As burlap, it has been used by fine artists as an alternative to canvas as a stretched painting surface. Sometimes, burlap is used as matting material for picture framing.

American glorious quilts from historical periods are important American folk art that reveals the national folklore and history in textile. Karey Bresenhan, the President of the Quilt Market and Festival in Houston, Texas writes: *"Most tobacco-sack quilts were made after the turn of the century, between 1910 and 1935. They were all hand-dyed, sometimes in pastels but more commonly in deep reds, blues, and browns. For the lighter colors it was necessary to bleach the bags first. Therefore, steps included ripping out the seams of the sack, washing, bleaching, dying, piecing and tying or quilting. Most of the quilts were pieced in some variation of the traditional Brickwork pattern to make the most efficient use of the shape of the sacks. Because the fabric is plain, embroidery is often used to decorate it."* There is a quilt from West Texas that iis embroidered with the brands of the region's ranches.

For example: the "Bull Durham" quilt. Top made of Bull Durham tobacco sacks that were dyed green and yellow and then pieced to make a design of concentric green frames on a yellow ground. Centers feature a sewn and painted bull standing on sewn grass with outline quilting.

Purchased in Bardstown, Kentucky, from a dealer who had purchased it in Mississippi tag read "Made from Bull Durham tobacco sacks." Each sack was hand-dyed before piecing.

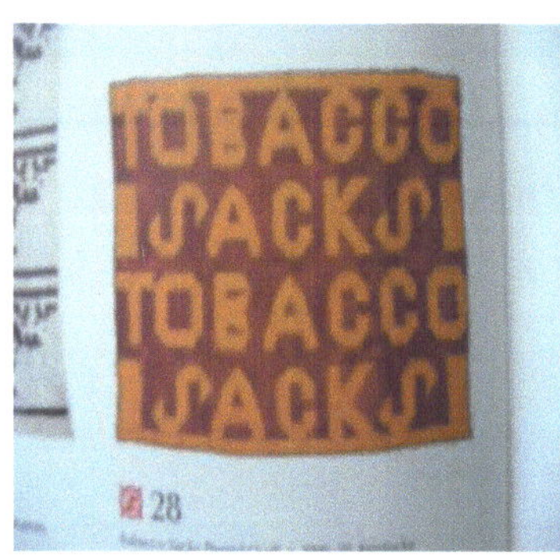

"Tobacco Sacks Pieced Quilt." c. 1900-1920. Kentucky.
Cotton and burlap tobacco sacks, 65 8 831/2"
Shelly Zegart's Quilts Collection, Louisville, Kentucky
Photographed by Steve Mitchell
It was suggested that this particular quilt was used to advertise tobacco sacks for sale – a patchwork billboard, so to speak.
(Duke & Harding, 1987)

Bible quotes: "When Mordecai perceived all that was done, Mordecai rent his clothes, and put on sackcloth with ashes, and went out into the midst of the city, and cried with loud and bitter cry; and came even before the king's gate: for none might enter into the king's gate clothed with sackcloth." It is difficult to say if he looked good in sackcloth from this quote, but it was unusual and perhaps brave for the culture of this time period, and, who knows, maybe he was setting a new fashion trend for his admires and followers.

Every fashion design student, who took the history of textile course, knows sufficient amount of information on major authentic resources such as cotton, linen, wool, but not so much about other minor organic man-made textiles. The use of burlap made from jute as a textile fabric is not a completely new idea. There are certain preparations for dressmaking and for obtaining the skills to make a dress.

Millinery and dressmaking constituted the higher end for female employment with the needle, as tailoring for men. Apprenticeship, as usually, started at age of fourteen and worked for two years prior to being considered experts in the field. The fabrics and supplies were expensive, and often imported. The training and learning required affordable textiles to exercise the trade skills.

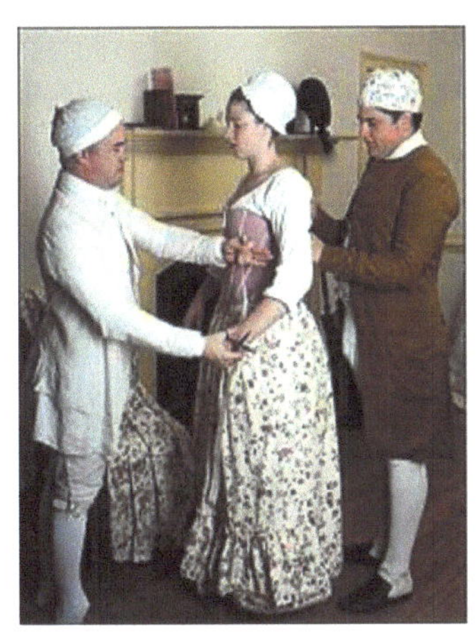

Toile de Jouy, sometimes abbreviated to simply "toile." In British or Australian terminology, a *toile* is a version of a garment made by a fashion designer or dressmaker to test a pattern. They are usually made in cheap material, as multiple 'toiles' may be made in the process of perfecting a design. *Toiles* are commonly called "muslins" in the United States, named for the cheap, unbleached cotton fabric available in different weights. *Toiles* were very popular during the Colonial Era in the United States and are highly associated with preservationist towns and historical areas, such as Colonial Williamsburg. When sewing clothing, a dressmaker may test the fit of a garment, using an inexpensive muslin or burlap fabric before cutting the intended expensive fabric, thereby avoiding potential costly mistakes.

"The tailor's customer ran the full social gamut – from the wealthy and elite to field slaves, and all the folks in between. The only discernible differences in the clothing for rich or poor was in the quality of the fabric. There was less difference in the quality of construction than in the type of fabric, since there was basically the same workmanship in a shirt for a slave as there was in a gentleman's fine silks. Most tailors did not sell fabric, so people selected fabric from a merchant in town and brought their own fabric to the tailor. Of the 16 or so tailor shops known to have been in Williamsburg in 1774, only two of those were merchant tailors – Prosser and Nicholson, located at opposite ends of the street." (Colonial Williamsburg Foundation, 2012) Tailors were almost always men. There is one reference in Williamsburg to a female tailor, who was paid for her work, but she never advertised as being a tailor, and no other reference was made to her, leading historians to surmise that she was likely the widow of a tailor who took up his trade when he died. (Tailor: Art of Cutting)

There are many pieces in fiction literature that describe seamstress trade of the Victorian era, about the 1840s. Please see the Charlotte Bronte's *Shirley,* the distressed seamstress appear in the work of Charles Kingsley *Alton Locke* and *Cheap Clothes and Nasty,* Charlotte Elizabeth Tonna *The Pearl of the Nation* and *The Wrongs of Woman,* Friedrich Engels *The Condition of the Working Class in England,* Thomas Carlyle *The Nigger Question,* and Charles Dickens *The Crimes,* among others.

Was burlap material used for clothing by the slaves? Historically, "*most field slaves were issued clothes at birth. When they were small, both genders usually wore plain cotton gowns. Right before winter they were issued a jacket and shoes. These must last them all year long. Most clothing was made of a burlap-like material called hemp. This was cheap and easy to find on the market. Slaves who worked in the big house were given hand-me-down clothes from the planter's family.*" (Clothing: Project by Students for Students, 2011)

So, everybody knows the statement "about being able to look good in a burlap sack" and in my opinion, it is a good title for this book. The inspiration came from different places and situations. For example, the unforgettable outfit from the episode of "I Love Lucy" TV show called "Lucy Gets a Paris Gown" where Lucy and Ethel get designers gowns made from burlap sacks and horse bags for hats. Classic!

"The girls absolutely LOVE their "Jacques Marcel designer dresses" and walk around like models . . .

Everyone stares at them, even the brilliant and acclaimed Jacques Marcel himself who eventually "copies" the designs, but by that time the girls have found out on the boys and burned the originals . . ."

The familiar phrase "you'd look good even in a potato sack" originated from a stunt pulled off by 20th Century Fox in 1952. Marilyn Monroe was under contract with the movie studio then and the studio wanted to showcase Monroe's beauty and prove that she didn't need to be dressed in gowns, furs and jewels to look good. The resulting photos – a series of shots of the beautiful 26 year old posing in heels and an Idaho Potato sack – are probably one of the most enduring photo sequences of the famous blonde.

There is more than one version of the story why and how Marilyn Monroe posed in a burlap potato sack. As the story goes, Marilyn was once chastised by a female newspaper columnist for wearing a low-cut red dress to a party at the Beverly Hills hotel. According to Marilyn, the columnist called her cheap and vulgar. Not stopping there, the writer then suggested that the actress would look better in a potato sack. The 20th Century Fox cinematography production studio decided to capitalize on the story by shooting some publicity stills of Marilyn in a form fitting burlap potato sack just to prove she would look sexy in anything. The photos were published in newspapers throughout the country. Another story was that someone just made an off-the-cuff statement that Marilyn could make a potato sack look sexy, and 20th Century Fox took the publicity stills to prove him right. Whatever story is behind, it is true to say that Marilyn always looked sexy, no matter what type of dress she wore.

Marilyn Monroe in a dress made of potato sack, with fringe! Circa 1951-1952

In current era, jute is not considered only as a crude material for gunny bags or Hessians mixed. Rather, it is mixed with cotton and silk to manufacture attractive range of decorative items, table mats, blankets, bags, wall hanging etc. The NIRJAFT (National Institute of Research on Jute and Allied Fiber Technology) in Kolkata, West Bengal, of India or as Europeans are saying Calcutta, acquired a unique technology to reduce the roughness of jute fiber and make it more malleable and soft.

Recently in India, one of the fashion show focused jute as a favorite material for fashion fanatics. The event has showcased large collection of jute apparels for both men and women. Many buyers, individuals and designers appreciated the texture and prints of the jute fabric. The Indian Ministry of textile promotes the exhibitions featuring garments and other crafted jute items. Some popular local designers like Pawan Aswani and Ashis Soni have accentuated jute in fashion shows like - Lakme India Fashion show 2000 and International Fairs in Dusseldorf, Germany. Fashion designer Prabha Mohanty is also among one who encourages use of jute as material for stylish garments. The latest fashion trend focuses on jute crafted accessories like jewelry, scarves, bags, ties, belts and many more. Even NIFT (National Institute of Fashion Technology) has participated in fashion shows in association with Institute of Jute Technology and India's premiere natural fabrics store located in Mumbai. The show patronizes jute creation to a large and high extent.

The positive aspect of jute is its durability and the soft silky natural golden color. Earlier, the fiber was presented in its most natural appeal. However to make it more soft and elegant, jute is mixed with cotton, wool, nylon, rayon, acrylic or polypropylene. Mixing jute with other material upgrades jute in terms of durability, feel, appearance, wash ability and resilience. The fashion world has been now revolutionized and modernized, and people are willing to accept the change. Jute can be beautifully ornamented by combining with bones, stones, metals and leather. Overall, jute, the natural vegetable fiber, has undergone a long way from the traditional sacks to high fashion world. The cost effective factor remains the positive aspect for manufacturers using jute as blending material for several fabrics.

- Jute apparels or garments are available for all occasions. Hand blocked, hand printed, embroidered, quilted and tie & dyed variety of jute garments look very fashionable. Apparel of jute can also be made with cotton blend called *juco*. Besides, jackets, sandals, clogs and different types of footwear are made to suit the exact needs of buyers. Jute apparels are available with a variety of designs and patterns. They look very fashionable and trendy thus lending an aesthetic appeal with a modern touch. These unique apparels help wearers to protect themselves from environment as jute material belongs to a pure ambit of organic and natural product. Now days, jute apparels are in great demand and used by top most famous designers to fabricate great looking formal and casual wear, accessories etc. Very fine threads of jute are also used by them to make false silk and related fit outs. There is a vast range of soft, long lasting and stylish jute apparels available in market in tremendous array of colors. Main features include – they can resist rough use, jute apparels fuse easily with any casual wear, cultural touch and finish makes it a treasured possession, and a long – lasting, too.

Jute jackets are very popular as they are made to be stylish and colorful. Fashionable footwear is made out of bleached and natural yarn-woven with burlap lining. Buttons, laces and other kinds of fashion accessories made out of jute yarn lend a rough yet chic look.

International jute fashion shows in India and Bangladesh in 2010 and 2011 witnessed beautiful and impressive jute products featuring the attractiveness of the diversified jute productions.

The jute fashion show organized by Jute Manufactures Development Council (JMDC), Ministry of Textile, and Government of India used the popularity of the Indian classical singer Rashid Khan who walked the ramp to promote jute as a fashion fabric together with Kolkata's top models offering collections from formal to *avant garde*, from classic to futuristic. Fashion designers today are creating great designs with this raw material. The best Indian haute couture is being crafted from it. The European fashion industry is also waking up to the possibility of using jute in apparel industry. The jute collection boasted the natural colors of the fabric with some were steeped in pastel hues. The textiles varied from strictly minimalistic to gorgeous machine embroidery and appliqué.

The integrity of burlap fabric qualities and its organic coloring inspired another TV show, the 2010 Project Runway Season 7 Episode 2 with a fundamental task to "create a party look from a potato sack." The challenge stems from the saying "she's so beautiful she'd even look good wearing a potato sack" was the motivation for young designers to design a party dress from burlap and "make it work" for their clients – very sophisticated young models. And they did, and in the end, the fashion designers made burlap look expensive.

The editor-in-chief of "Harper's Bazaar" magazine, Glenda Bailey, says "I really like all the neutral color and natural materials designers are using, especially the linens and burlap. White has been such a big story this season and gray is, as well. After a fall season with so much black in it, it's really no surprise that in the new spring season gray is the new black." Cynthia Rowley (1958 -), American fashion designer from the West Village of New York city, continues the fashion conversation by adding her opinion on "mixing everything" in fashion: "So, it's like, there's maybe like this linen burlap jacket car coat with a little chiffon dress underneath. So it's like mixing really true evening pieces, like little dance dresses, with something that's much more rugged, as a contrast."

Designer Miu Miu made burlap dresses punchy for Spring Collection 2009 by introducing 'burlap graffiti fashion.' We would assume that the look and the burlap fabric were chosen for the times of recession, but the price says differently with dresses cost at $1,500, purses costing around $1,400 and shoes cost at $1,600.

Trust Donna Karan to make burlap look effortless and sophisticated. Featuring matte jerseys, light linens, and the aforementioned burlap – that somehow looked weightless – the Donna

Karan's Spring-2010 ready-to-wear collection was all about draping with an overall upswept air. According to her show notes, the line was "inspired by the power of the elements – the sky, water, wind, sand, earth, and fire."

May is a Potato Lover's Month in Idaho. It's time for the Idaho Potato Commission's (IPC) the 21st annual Potato Lover's Month Retail Display Contest and roll up that burlap carpet for winners.

Use your imagination to picture the burlap potato sack obsession!

Burlap sacks are the primary containers of most farm produce like potatoes and coffee beans. This is because they are categorized as "food grade" containers, or those that are safe and non-hazardous and will be used with products that are going to be made into food. As you may or may not know, burlap sacks are used in a variety of different ways. For example, in some farm areas, burlap sacks are used to store natural fertilizers. And by natural fertilizers, we mean animal waste. This demonstrates why burlap sacks are such a versatile option for carrying items other than just food products.

Another reason why burlap sacks are best for carrying harvested products is because they are gentle. They are made from Jute, a special type of woven fabric that is soft just enough to not damage whatever the sacks contain. And although they are gentle, they are also undeniably very durable. Just imagine how much weight one sack can contain. And no matter how heavy their load is, they never disintegrate!

And just like we mentioned in the beginning, burlap sacks are food grade containers. So, when you happen to get a sack to use at home and you suddenly can't find a place to store your food supplies, then burlap sacks are available to solve your problem.

Families with children encourage them to use burlap sacks as means of entertainment, rather than use electronic devices such as the computer and various gaming consoles. Conduct a mini sack race with kids, and whoever wins, can get a prize. This way, not only use burlap sacks are keeping children physically active, but also they are pulled away from the negative effects of technological devices. People find that there are many ways to use burlap sacks. The use of imagination and practical instincts will do the originality. And just by using burlap sacks in everyday life, people become a part of the community that is saving our environment.

The fact that burlap sacks are mostly recycled appeal to a lot of people as well. Given that we live in a time when global warming and various climate problems are sprouting here and there, everyone just wants to do their part in helping the planet. Of course, with burlap sacks that are turned into bags, anyone can do just this. By using burlap bags, the use of plastic bags and other non-biodegradable containers will come to an end.

On the left: 40-inch-5-oz natural burlap fabric

"IN THE LONG AGO TIME WHEN THINGS WERE SAVED,
WHEN ROADS WERE GRAVELED AND BARRELS WERE STAVED.

WHEN WORN-OUT CLOTHING WAS USED AS RAGS,
AND THERE WERE NO PLASTIC WRAP OR BAGS.

AND THE WELL AND THE PUMP WERE WAY OUT BACK,
A VERSATILE ITEM, WAS THE FLOUR SACK."

("The Flour Sack" by Colleen Hubert)

Recycling was a way of life and a matter of necessity in American culture. On the last weekend of each April, Michael Zahs, a history teacher and a collector, puts out for the public almost thousand feed sacks and examples of the guilt and clothing made from them. As a fact, the historic true is that in the past in every household every bit of fabric, every coverlet, apron, diaper, and bonnet – was first a container for sugar, animal feet or agricultural seed. For centuries, countless items were transported in bulky wooden barrels and boxes; and they were heavy to carry and store. In the mid-1850s an overabundance of burlap fabric made bags cheap to produce and improved dramatically with the use of sewing machine in household. From transportation and storage perspective, to ship goods in bags were as good as in barrels, included flour, sugar, seed, animal feed, ham and sausages, and even ballots. At a time when many rural families had limited recourses, these bags were considered nearly as valuable as the items they contained.

Left: Two women in feed sack dresses (*National Geographic*, 1947).

Right: Instructions from a chicken feed sack.

"TO WASH OUT INK" instructed how to remove the logo or the label from the flour sack that was not an easy task. Combination of washing and soaking included lye, lard, Fells-Neptha soap and bleach. Thrifty women used the whitened textiles to stitch young children clothing, curtains, bed sheets, towels, and even husband's underwear. The feed sack instruction to wash out ink read: *"Soak overnight in soap suds and cold water then wash thoroughly in warm soap suds until ink has been loosened. Rinse well and if necessary boil for ten minutes to restore the natural whiteness of the cloth."*

Over time, manufacturers realized increasing value of the bags could increase the profits. Bags were stamped with stitching lines for reuse as household items and with embroidery patterns classic slogans series such as "Wash on Monday," "Iron on Tuesday," "Bake on Wednesday." Producers wonted ambitious women as consumers motivate their husbands to buy more animal feet or any other produce in order to complete the entire set of textile with advertising labels. Instead of printing on the sack, factories affixed their logos on easily removable paper labels.

- Historic fact: In the mid-20th, mills started producing sacks in printed fabrics. 40 mills made fabric for bags in thousands of different patterns.

Only a woman that is wearing sixe 2 would use a two-feed sack for a dress, a typical woman's dress took three-feed sack. Wives and daughters instructed husbands and fathers to purchase feed in sacks in particular patterns so that they could complete dresses. There were floral, border prints (used for pillow cases and curtains) and children prints with animals and cowboys. If the pattern sold well, it would make the manufacturers reproduced them in textile yardage for better sales.

- During depression era plaids and stripes were more limited and solid colors dominated
- During the wartime era of the 1940s, the patterns were patriotic kind and prints with "V" for "victory" and Morse code appeared
- In the 1950s, there were "exotic" Mexican and tropical motives in fabrics, and it was the start of the growing popularity of Mickey Mouse

Technological progress in the industry during the World War II, brought changes and half of the items that previously sold in cloth bags made from burlap or cotton were supplemented by brown paper or plastic as it was cheaper to produce and considered more sanitary. Over the next ten to fifteen years, cloth bags disappeared. There are left the small mills operated for tourist industry and in the Amish and Mennonite communities. Seamstresses stitched quilts from tiny scrap with enthusiasm and lure. The challenge to make something new by using every bit is a creative process itself. "Recycling idea is far from new and trendy – it is a way of life style that helped survive economically countless generations." (Linzee Kull McCray)

In the modern time, use of the vintage and reproduction of feed sacks has a real "shabby chic." Vintage feed sacks from cotton bags used for flour and sugar as well as the burlap bags used for animal feed in variety of prints temp buyers to collect enough matching fabric to make dresses or curtains for the kitchen. Feed sacks continue to be very collectible "fabric from the farms."

In 1984, Professor Tom Trusky of the English Department at Boise State solicited Idaho potato companies for samples of their burlap sacks-to serve as inspiration for the graphic artists illustrating the 1985 "All Idaho" issue of *Cold-drill,* the university's literary magazine. Those potato sacks were put on display at Boise State in 1985 (in conjunction with the publication of the magazine) and again in 1995 (as part of an exhibit entitled "Sack Art.") Following the 1995 exhibition, the sacks were transferred to the University Archives in Albertsons Library, where they became part of Record Group 425, Records of *Cold-drill.*

They resurrected twenty-four of the sacks weighing 15-pounds and put them on display at the Albertsons Library during January and February 2006. Each sack measures approximately 22 by 36 inches and, when full, can hold 100 pounds of potatoes. Once the predominant form of packaging used by distributors to ship their potatoes to market, the burlap potato sack is an important cultural symbol in potato-growing Idaho. The images on the sacks tell many stories of interest to students of advertising, marketing, the graphic arts, and social and cultural history. While burlap sacks are still used by Idaho's potato distributors, their supremacy is now challenged by mesh, paper, and poly bags. Altogether, there are 105 sacks from Idaho and other Western states in the collection.

This burlap drawstring bags measure 5×6 and are embellished with a wood heart which has been hand stamped with "love." These are a perfect way to say 'thank you' to any guests at any party. They will look great in any reception setting that features rustic, vintage or barn style themes. In addition to their vintage inspired style, these burlap bags also have functional value, because they're a great choice to use as candy buffet bags! Not only will your guests enjoy all the sweet treats but will have a sweet keepsake bag to remember the special day!

Below is a handmade crochet bag from jute industrial rope used as a yarn decorated with green flower, c.2010, courtesy of Mrs. Ida Tomshinsky.

"You can hug a tree or a tree can embrace you." (Ida Tomshinsky)

There are plenty international establishments and strong market goodwill toward a wide client base for manufacturing and export of fine quality jute and canvas bags like: designers jute bags, jute shopping bags, and canvas beach bags. There are wine bottle jute bags, sack bags, sling bags, Christmas bags, travel bags, etc. For example, H. A. Export has established from late 90's a comprehensive infrastructure leaded by expert craftsmen. Their innovative trendy produce delivering capacity is about 50,000 jute bags and 50,000 canvas bags. Oasis is another leading manufacturer of wholesale and exporter of designer jute bags and pouches. Established in 2006, and since then creating a niche of jute and cotton products, Green Packaging Industry Pvt. Ltd. manufactures and export superior quality jute merchandize and similar products such as ladies jute bags, fashion jute bags, and jute tote bags.

Ethnic/Earth fashion trend includes natural trims jute, linen, stones, wood beads, and embroidery.

Earth tones, drawstrings, long day dresses, and so called 'earthy' traditional prints. Prada's real fantasy includes Prada orange striped canvas platform jute and rubber wedges.

Jute is identified primarily with old-fashioned sacks for the cement and sugar industries. But with better designs, a rising environmental concern and new marketing strategies, the Indian jute industry is marketing a range of everyday products made with the 'golden fiber' of which the country is the largest producer, consumer and exporter.

The industry, which in the past mainly focused on packaging sacks used by the cement and sugar industries, is gearing up for a complete makeover, with a range of designer merchandise, including school bags carried by millions of children. "You can carry them with style," said Aditi Shukla, a student of fashion design in the national capital. "These green bags will be trend - setters soon."

There are also socio-economic issues associated with jute. It supports four million farm families, mainly in the eastern states, notably West Bengal, and provides direct jobs to 260,000 industrial workers and 140,000 people in tertiary activities. According to the council, India is the world's largest producer of jute-based goods, averaging 1.6 million tons per annum in the past five years with domestic sales of 1.4 million tones and exports of 285,000 tones.

Environmentalists believe that plastic bags are destroying our environment. Even biodegradable plastics degrade very slowly with proper conditions such as exposure to sunlight, oxygen, water, etc. The degradation process of biodegradable plastic releases CO_2 (Carbon Dioxide) which is considered as a GHG (Green House Gas) and contributes to global warming. In order to leave a better climate for future generations, the use of plastic must be avoided, no matter whether it is biodegradable or recyclable. Jute fibers are vegetable fibers which grow naturally. Fabrics or shopping bags made of jute are 100% biodegradable and the best replacement for plastic bags.

Eco-oriented clothes mean a wardrobe carefully put together with organic natural materials from organic cotton, linen, jute, alpaca, merino, or other earthy textile. The use of low-impact dyes or none at all; instead, the materials are left natural white, brown, tan, gray or black. Not only is the simplicity beautiful – it keeps workers from being exposed to toxic dyes. Eco-friendly fabrics are produced adhering to strict fair trade standards. That means that the designs are coming straight from local artisans in places like Ecuador, India, Bangladesh, Pakistan, Guatemala and Peru, and the artisans benefit directly. The merge of the smart designs with quality perspective manages the fashion forward and maintain ethical principles making high-fashion designs eco-green far beyond the granola crowd. In the end, these clothes are irresistible to the eco-savvy shopper. As the folk-rock group "Burlap to Cashmere" reclaims, 'burlap to cashmere' is in all: the turf in top form with an inspired blend Mediterranean rhythms, rootsy textures, and high-knit harmonies, which lend the unmistakable fresh air of earth.

People say that you can be beautiful in a potato sack: *"Looks like a sack, feels like a sack, is a sack,"* and as one of 60's Pop Art potato sack dress says: *"Fill with 100 lbs or more of charm, and save money on new French inspired creations."*

Original letter from a supplier:

10-Burlap Coffee Bags Bag Sack Jute ARCO Race: $18

THESE ARE USED, WILL NOT LET ME CHOOSE USED ON THE CONDITION

Hi, you are offered on a bundle of 10 burlap coffee bags. Each bag is in excellent condition. A number of the bags are from Brazil, there are airbags from other countries as well.

You will receive repeats.

The photo is of some of the bags we have roasted. ARCO coffee roasts fine coffee from around the world so the bags you actually receive may not be the ones pictured. I try to make each bundle as interesting as my stock allows.

All the bags are quite strong; they come to us with 130# to 150# of green coffee in them. The bags are empty now, of course. Each bag measures about 2 feet X 3 feet.

Use the bags as decoration in your home or shop, in the garden as a tree wrap, to protect new grass seeds, store tools, gunny sack races, or as a really neat wrapping paper.

Buyer pays shipping; I get a discount on shipping with UPS which pass on to you. Please use the shipping calculator at the bottom right corner of this listing to get the shipping cost. Each bundle of 10 bags weighs between 20 and 26 pound, I only charge for 20 pound.

PLEASE VISIT OUR STORE

IF YOU WIN OUR BURLAP BAGS, I WILL THROW IN FREE SHIPPING FOR UP TO 10 BAGS OF COFFEE FROM OUR STORE.

THANK YOU AND GOOD LUCK.

Bibliography

Blank, Nayne. *Sack.* – *DailyBibleStudy.com, 2005. [Sack definition and more]*

Bresenhan, Karoline Patterson, and Nancy O'Bryan Puentes. *Lone Stars: A Legacy of Texas Quilts, 1836-1936.* – Austin: University of Texas Press, 1986.

Carter, H. R. *Modern Flax, Hemp and Jute Spinning and Twisting: A Practical Handbook for the Use of Flax, Hemp, and Jute Spinners, Thread, Twine and Rope Makers.* – Nabu Press, 2010. – 232 pages.

Dennis, Morgan. *Burlap.* – E. M. Hale, 1945.

Donenfeld, Maya. *Reinvention: Sewing with Rescued Materials.* – Wiley, 2012. – 192 pages.

Duke, Dennis, and Deborah Harding. *America's Glorious Quilts:* Beaux Arts Editions. – Hugh Lauter Levin Associates, Inc., 1987.

Enderlen-Debuisson, Marie and Caroline Laisne. *Chic Bags: 22 Handbags, Purses, Totes, and Accessories to Make.* – New York: St. Martin's Griffin, 2006. – 160 p.: ill.

Smart carryall from burlap fabric

Fressard, M. J. *Creating with Burlap: Decorating, Painting, Embroidering.* – Sterling, 1970.

"Green Craft Magazine: Pillows and Burlap for the Home." – Autumn, 2010.

"Jute and Jute Fabrics: Magazine" – Bangladesh: Bangladesh Jute Research Institute. – 12 issues/12 months.

Jute: Webster's Timeline History, 1800 BC-2007. – ICON Group International, Inc. – ICON Group International, Inc., 2009. – 66 pages.

Lee, Janet. *Living in a Nutshell: Posh and Portable Decorating Ideas for Small Spaces.* – Harper Design, 2012. – 208 pages.

Lefrance, Emily. *Culture and Manufacture of Ramie and Jute in the United States.* – Nabu Press, 2010. – 40 pages.

Lewis, J. C., eHow contributor. – eHow.com, January 21, 2012.

History of Burlap I eHow.com http://www.ehow.com/about_4577712_history-of-burlap.html#ixzz1jjBIVj1y

What Is Burlap Fabric? I eHow.com http://www.ehow.com/about_6591726_burlap-fabric_.html#ixzz1jjCnkL5P

What Is Burlap Made Of? I eHow.com http://www.ehow.com/how-does_4600964_what-burlap-made.html#ixzz1jjC5UtGL

Londo Great Britain. *Indian Trade Enquiry: Reports on Jute and Silk.* – Nabu Press, 2010. –104 pages.

Love, Herald, T. *Burlap.* – Authorhouse, 2004. – 296 pages.

McGinnis, Eddy. *Feedsacks! Beautiful Quilts from Humble Beginners.* – Kansas City: Star Book, 2006. – 126 pages.

Miller, Susan. *Vintage Feed Sacks: Fabric from the Farm.* – Schiffer Publication Limited, 2007. – 160 pages.

Nixon, Gloria. *Feedsack Secrets: Fashion from Hard Times.* – Kansas City: Star Books, 2010. – 144 pages.

Paull, Charles H. *English Lessons for the Jute Industry.* – Nabu Press, 2010. – 90 pages.

Sommer, Elyse. *Make It with Burlap.* – Lothrop, Lee & Shepard Co., 1973. – 96 pages.

Sons, Dennis. *Burlap: A Mainstream Fabric.* – Wryte Stuff.com, 2009.

Stapel, Jane Clark. *A Brief History of 'Feedsacks.'* – www.planetpatchwork.com/feedsack.htm

Tomlison, Jim, Carlo Morelli and Valerie Wright. *The Decline of the Jute: Managing Industrial Change.* – Pickering & Chatto Ltd, 2011. – 219 pages.

Wingate, Isabel Barnum. *Textile Fabrics and Their Selection*: 8th/ed. – Englewood Cliffs: Prentice-Hall, Inc. – 1984. – 704 pages.

Woodhouse, Thomas and Thomas Milne. *Jute and Linen Weaving.* – Nabu Press, 2010. - 630 pages.

Woodhouse, Thomas. *The Jute Industry: from Seed to Finished Cloth.* – Tredition: 2011. – 236 pages. (Tredition Classic Series)

http://www.quotesea.com

Children Literature

Hardy, Mary Smith. *Uncle Hubbard and the Burlap Sack.* – iUniverse, 2012. – 48 pages.

Left: Jute rug outdoors; Right: A close look on the weaving

"Two roads diverged in a wood, and I,
I took the one less traveled by,
And that has made all the difference."

(Robert Frost)

www.ingramcontent.com/pod-product-compliance
Ingram Content Group UK Ltd.
Pitfield, Milton Keynes, MK11 3LW, UK
UKHW060122300726
14090UKWH00002B/309

* 9 7 8 1 4 7 9 7 6 1 2 0 3 *